The Number System

BJC MATHEMATICS GUIDE

The Number System: BJC Mathematics Guide by Zemi Stewart, 2018

This book or parts thereof may not be reproduced in any form, stored in a retrieval system, or transmitted in any form by any means—electronic, mechanical, photocopy, recording, or otherwise—without prior written permission of the author.

All inquiries regarding this publication, ordering information, requests for usage rights or corrections should be sent by email to mathwithmisszemi@gmail.com.

While the author has made every effort to provide accurate Internet addresses at the time of publication, neither the publisher nor the author assumes responsibility for errors or changes that occur after publication.

ISBN-13: 978 1 7259 0212 1
ISBN-10: 1725902125

Published in Nassau, Bahamas
Printed and bound in the United States of America by CreateSpace

Table of Contents

Welcome	5
About the BJC	6
BJC Syllabus	7
Expectations	8
Introduction	8
The Real Number System	**9**
Key Vocabulary	9
Exercise 1	10
Whole Numbers	**11**
Exercise 2	11
Even & Odd Numbers	12
Prime & Composite Numbers	12
Factors & Multiples	13
Exercise 3	14
Prime Factorization	15
Exercise 4	15
Highest Common Factor	15
Lowest Common Multiple	16
Using Prime Factors to Find HCF & LCM	16
Exercise 5	17

Exponents & Roots — 18

    Exercise 6 — 19
    Laws of Exponents — 20
    Exercise 7 — 20
    Square Roots — 21
    Exercise 8 — 22
    Cube Roots — 23
    Exercise 9 — 23
    Standard Index Form — 24

The Four Rules — 25

    Addition & Subtraction — 25
    Exercise 10 — 25
    Multiplication & Division — 26
    Exercise 11 — 26
    Order of Operations — 27
    Exercise 12 — 28

Number Patterns & Sequences — 29

    Exercise 13 — 30

Exam Style Questions — 31

Resources — 39

Welcome

Hi there,

Welcome to this short guide introducing the real number system and the study of whole numbers. Here, we explore the forms used to write numbers, standard mathematics vocabulary (factor, multiple, even, odd, etc.), highest common factors, lowest common multiples, prime factors, the four rules, powers, roots and number patterns and sequences. This guide provides easy to read explanations as well as exam style questions and additional resources for further study.

In my opinion, math should be both studied and practiced. I encourage you to actually *read* this book, much like you would your other text books. Don't just skip along to the questions, actually read to *learn* about Mathematics: its rules and principles.

I hope you enjoy this guide and the many others in this series. It was inspired by you!

Happy learning!

-Zemi Stewart,
Founder, Math with Miss Zemi

About the BJC

The Bahamas Junior Certificate (BJC) is an examination curriculum designed by the Bahamas' Ministry of Education. The examination is usually undertaken in grade 9 in public secondary schools however, at private institutions, anyone over the age of 16 is eligible to write the examination. The Mathematics examination is divided into two (2) papers of varying difficulty:

BJC Syllabus

- ✓ Whole Numbers
- ✓ Number Patterns & Sequences
- ✓ Decimals, Fractions and Percentages
- ✓ Ordering (Comparing Quantities)
- ✓ The Four Rules
- ✓ Directed Numbers
- ✓ Ratio & Proportion
- ✓ Estimation/Approximation
- ✓ Sets & Venn Diagrams
- ✓ Measures
- ✓ Time
- ✓ Average Speed (Speed, Distance, Time)
- ✓ Foreign Currency & Exchange Rates
- ✓ Consumer Mathematics
- ✓ Symmetry
- ✓ Area, Perimeter & Volume
- ✓ Statistics
- ✓ Probability
- ✓ Inequalities
- ✓ Expressions & Formulae
- ✓ Graphs
- ✓ Cartesian Coordinates
- ✓ Geometrical Terms
- ✓ Triangles, Quadrilaterals and Circles
- ✓ Measurement of Lines & Angles
- ✓ Constructions
- ✓ Transformations

Expectations

The examiners expect you to be able to:
- Read and write numbers in exponential form
- Know the symbols indicating square and cubed roots
- Calculate roots up to 400 without a calculator
- Calculate cube roots up to 1,000 without a calculator
- The four rules applied to whole numbers
- Recognize simple number patterns, e.g. squares, cubes, triangular numbers, etc.

Introduction

An understanding of the number system is essential for the study of Mathematics. Though whole numbers are a key part of the BJC examination syllabus, these represent just a small subset of the entire number system. I will therefore begin our study of whole numbers by introducing you to the Real Number System.

> Note: Real numbers are actually a subset of a wider group of numbers called "complex numbers". Complex numbers have a real and an imaginary part; however, these are beyond the scope of this course.

The Real Number System

At this point I know you must be wondering, "Why are they called real numbers?" Well, what we know today as real numbers actually had no name before imaginary numbers were thought of. They were given the name "real numbers" only because they were not imaginary. (Again, we will not discuss complex and imaginary numbers in this text nor will you see them at the BJC level.)

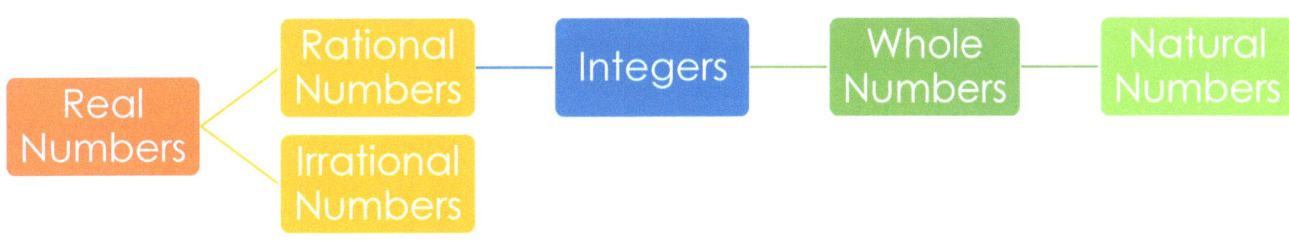

Figure 1 The Real Number System

As we see above, real numbers are split into two groups: rational numbers and irrational numbers. Rational numbers are further split into integers, whole numbers and then natural numbers.

Key Vocabulary

real numbers: all of the rational and irrational numbers

irrational numbers: numbers that cannot be written as fractions

When we try to write irrational numbers as decimals, the digits go on forever, with no recurring (or repeating) digits.

For example, the number π (pronounced "pi") is considered irrational because the numbers go on forever with no repeating digits: π = 3.1415926535…

> When I attended university, my classmates would often quiz each other on who knew the most digits of π. I knew about 4 digits, but some knew 100s. I am just noting this here in case you want to impress your peers by learning a few extra digits of π in advance.

rational numbers: a number than can be made by dividing two integers

That is, rational numbers can be written as a fraction where both the numerator and denominator are integers.

Examples: $\frac{1}{2}, \frac{5}{8}, \frac{35}{5}$, etc.

7 is a rational number ($\frac{7}{1}$); 0.75 is a rational number ($\frac{3}{4}$); etc.

What other rational numbers can you think of?

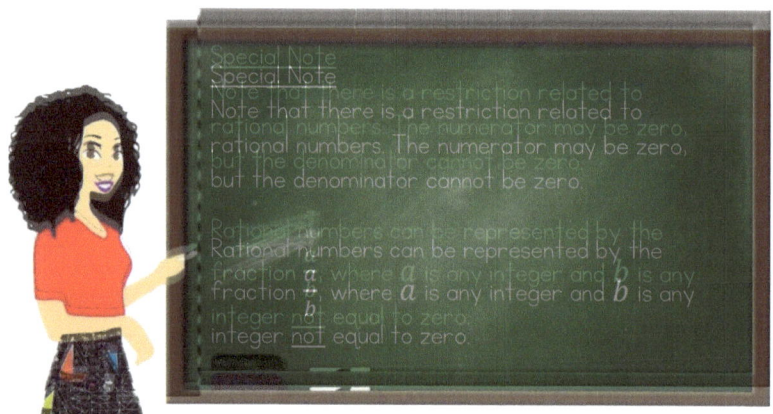

Special Note

Note that there is a restriction related to rational numbers. The numerator may be zero, but the denominator cannot be zero.

Rational numbers can be represented by the fraction $\frac{a}{b}$, where a is any integer and b is any integer not equal to zero.

integers: integers are whole numbers, both positive and negative, including zero

These numbers are ...-3, -2, -1, 0, 1, 2, 3...

Positive integers are integers greater than zero (+1, +2, +3, +4, and so on). We write these simply as 1, 2, 3, 4, etc.

Negative integers are integers less than zero (-10, -9, -8, and so on).

whole numbers: positive integers and zero

Examples are 0, 1, 2, 3, 4, 5, 6....and so on, and so on.

There are an infinite number of whole numbers. We therefore say that the series of whole numbers is infinite.

Exercise 1

Natural or **counting numbers** are another element of the real number system. Write the definition of a natural number below including examples. Ask your teacher if he or she considers the number "0" to be part of the set of natural numbers.

Whole Numbers

Numbers, and particularly whole numbers, play a huge role in our day-to-day lives. When you dial your best friend's number, you use whole numbers. When you note the date or the year, you use whole numbers. When you count, you use whole numbers. Whether you like math, love math or totally despise math, numbers are in fact all around you and they have been since you were born.

Whole numbers are typically written in **standard notation** or **standard form** using the first ten digits 0, 1, 2, 3, 4, 5, 6, 7, 8, 9 to present the number. For example, 34,125.

Whole numbers can also be expressed in expanded form and word form. The **expanded form** of a number breaks it down into parts. For example, 34,125 in expanded form is:

$$30 \text{ thousand} + 4 \text{ thousand} + 1 \text{ hundred} + 2 \text{ tens} + 5 \text{ ones}$$

or

$$30{,}000 + 4{,}000 + 100 + 20 + 5$$

When we say a number out loud, we recite the number in **word form**. For example, 34,125 in word form is:

"thirty-four thousand, one hundred twenty-five".

Exercise 2

1. Add five hundred and six, one hundred and ninety-one, and twenty-nine.

2. Find the value of the following sums:
   (a) $30 + 87 + 102$
   (b) $234 + 62 + 36$

3. In a church youth group, there are 42 attendees. 8 of the attendees are boys. How many are girls?

4. Write $600 + 30 + 1$ in **standard form**.

Even & Odd Numbers

Any number that can be divided by 2 is called an **even number**. If a number is not divisible by 2, it is called an **odd number**.

In all cases,
- If a number ends with an even digit, it is even.
- If a number ends with an odd digit, it is odd.

Examples of even numbers are 2, 4, 6, 8, etc.
Examples of odd numbers are 1, 3, 5, 7, 9, etc.

Prime & Composite Numbers

A **prime number** is a number divisible only by 1 and itself.

The first 20 prime numbers are:
2, 3, 5, 7, 11, 13, 17, 19, 23, 29, 31, 37, 41, 43, 47, 53, 57, 61, 67, and 71

Many are often surprised that 2 is a prime number. 2 is divisible only by 1 and itself and is therefore considered prime. It is in fact the only even prime number.

Q. Why is 2 the only even prime number?
A. All other even numbers are divisible by 2.

For example, the divisors or factors of **4** are **1, 2** and **4**. Therefore, 4 is not prime. It is instead a **composite number**.

A **composite number** is a whole number that can be divided exactly by numbers other than 1 or itself.

For reasons beyond the scope of this course, 1 is not considered to be a prime number.

Prime numbers are quite significant in mathematics as every composite number greater than 1 can be expressed as a product of prime numbers. This fact is known as **the Fundamental Theorem of Arithmetic**, but is quite often referred to as **Prime Factorization**. We will explore Prime Factorization after discussing factors and multiples in the next section.

Factors & Multiples

The **factors** of a number are the numbers that divide into it exactly, with no remainder.

Examples
(1) List the factors of 36.
The factors of **36** are **1, 2, 3, 4, 6, 9, 12, 18,** and **36**.

(2) List the factors of 17. Is 17 prime or composite?
The factors of **17** are **1** and **17**. Therefore, 17 is a prime number.

In order to find the factors of a number we normally think in pairs. In the example above the factor pairs are:

1 and 36 → because 1 x 36 = 36
2 and 18 → because 2 x 18 = 36
3 and 12 → because 3 x 12 = 36
4 and 9 → because 4 x 9 = 36

The factor pairs of 16 are:
- 1 and 16
- 2 and 8
- 4 and 4

Q. List the factor pairs of 30.

A **multiple** is the result of multiplying a number by an integer.

For example, the multiples of **3** are **3, 6, 9, 12, 15, 18,** and so on.

Here are some key tips to remember about multiples:
- The first multiple of a number is the number itself.
- There are infinitely many multiples of a number.
- When you think multiples, think times tables.

Q. List the first five multiples of 8.

Exercise 3

1: Here is a list of 9 numbers: 3; 4; 7; 8; 9; 10; 12; 16; 17

From the list, write down:

(a) two even numbers

(b) two numbers with a sum of 18

(c) a prime number

(d) a factor of 12

(e) a multiple of 5

2: Jami says 27 is a prime number. Explain why Jami is wrong. Respond in full sentences.

3: List all of the factors of the following numbers:
   a: 2

   b: 24

4: List the first 8 multiples of 4:

5: In the box below, circle all of the factors of 48:

22		13	6		10		12		
	9			8		5		11	
3		4		7	29		2		19

Prime Factorization

We often find it useful in mathematics to write a number as a product of its prime factors. In fact, we learned earlier that every composite number greater than 1 can be written as a product of prime numbers. These prime numbers are the number's prime factors.

We often depict a number's prime factorization using factor trees or branches.

For example, the prime factorization of 36 is:

OR

Note:
- No matter which factors of the number you choose, the resulting prime factors should be the same. (In the example above, we ended up with same prime factors in both instances.)
- If you multiply the prime factors, you should end up with the original number. (That is, the product 2 x 2 x 3 x 3 should equal 36.)
- 1 is never used within a factor tree.

Q: Write 45 as a product of its prime factors.
A: 45 as a product of its prime factors is 3 x 3 x 5.

Exercise 4

Go to https://www.transum.org/Maths/Activity/Prime/ and complete 10 prime factor trees.

Highest Common Factor (HCF)

The **Highest Common Factor (HCF)** of two numbers is the highest number that divides exactly into both numbers.

Example
Q: What is the highest common factor of 24 and 40?
The factors of 24 are 1, 2, 3, 4, 6, 8, 12 and 24.
The factors of 40 are 1, 2, 4, 5, 8, 10, 20, and 40.

We note that 8 is the highest common factor between the two numbers and, therefore, 8 is the highest common factor of 24 and 40.

Lowest Common Multiple (LCM)

The **Lowest Common Multiple (LCM)** of two numbers is the lowest number that is a multiple of both numbers. The LCM may also be interpreted as the lowest number that appears in the times tables of both of the numbers.

Example
Q. What is the lowest common multiple of 24 and 40?
The multiples of 24 are 24, 48, 72, 96, 120,…
The multiples of 40 are 40, 80, 120, 160, 200…

120 is a common multiple of 24 and 40 and appears to be the lowest common multiple. Therefore, 120 is the lowest common multiple of 24 and 40.

Using Prime Factors to find the HCF & LCM

You can find the HCF and LCM of two (or more) numbers by listing all of the factors or a few of the multiples like we did in the previous examples; however, for large numbers this may prove both difficult and time consuming. Instead, you can use the prime factors. We will explore this method below.

Example
Q. Find the LCM and HCF of 24 and 40.

Follow the steps below:
1. Find the factor trees of the two numbers.

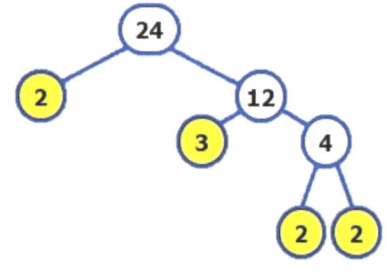 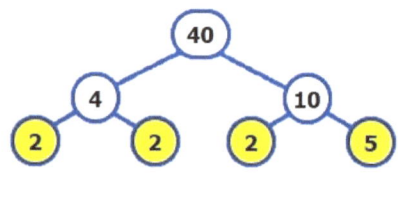

2. Write the factors on top of each other:
   24 = 2 x 2 x 2 x 3
   40 = 2 x 2 x 2 x 5

3. Highlight or circle numbers that appear in both rows.
   24 = 2 x 2 x 2 x 3
   40 = 2 x 2 x 2 x 5

To get the HCF, multiply the numbers that are in both rows (that is, the numbers you have highlighted).

The HCF of 24 and 40 is therefore 2 x 2 x 2 = 8.

To get the LCM, multiply each number you see.

The LCM of 24 and 40 is therefore 2 x 2 x 2 x 3 x 5 = 120.

> This may seem like quite a bit of work, but if you have very large numbers, this method is very helpful.

Exercise 5

1. Find the highest common factor of the following pairs of numbers.
    a. 32 and 36

    b. 40 and 8

2. Find the lowest common multiple of the following pairs of numbers.
    a. 3 and 8

    b. 12 and 20

3. Find the highest common factor and lowest common multiple of 52 and 18 using the prime factors method.

Exponents & Roots

An **exponent**—also referred to as the "power" or "index"—is a small number placed to the upper-right of a number which signifies how many times the base number is multiplied by itself.

For example,

$$5^4 = 5 \times 5 \times 5 \times 5 = 625$$

with the exponent being 4 and the base being 5.

The above example reads "5 to the power of 4" and is simply 5 multiplied by itself 4 times. 5 is the **base** and the **exponent** or **power** or **index** is 4.

In **index form** or **exponential form**, $4 \times 4 \times 4 \times 4 \times 4 \times 3 \times 3 \times 2 = 4^5 \times 3^2 \times 2$

When a number is multiplied by itself twice, we call it a **square number**.

More generally, a square number is a^2, where a is an integer.

Example, 27^2 reads "27 squared" or "27 to the power of 2".

The first 10 square numbers are: 1, 4, 9, 16, 25, 36, 49, 64, 81, and 100.

Because...
$1^2 = 1 \times 1 = 1$ $4^2 = 4 \times 4 = 16$ $7^2 = 7 \times 7 = 49$ $10^2 = 10 \times 10 = 100$
$2^2 = 2 \times 2 = 4$ $5^2 = 5 \times 5 = 25$ $8^2 = 8 \times 8 = 64$
$3^2 = 3 \times 3 = 9$ $6^2 = 6 \times 6 = 36$ $9^2 = 9 \times 9 = 81$

When a number is multiplied by itself three times, it is called a **cube number**.

More generally, a cube number is a^3, where a is an integer.

The first 5 cube numbers are displayed below.

Exponent or Index Form	Expanded Form	Value
1^3	$1 \times 1 \times 1$	1
2^3	$2 \times 2 \times 2$	8
3^3	$3 \times 3 \times 3$	27
4^3	$4 \times 4 \times 4$	64
5^3	$5 \times 5 \times 5$	125

Q: Why are they called square numbers?

Square and cube numbers get their names because of the patterns they form.

For example, we note that by counting the dots that form a square pattern, we end up with **square numbers**:

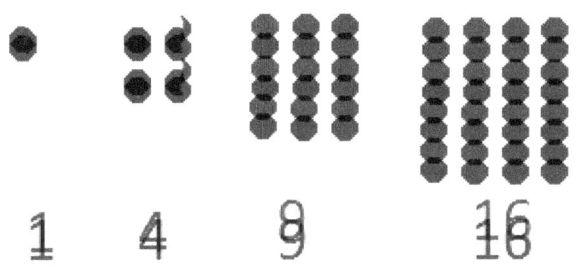

1   4   9   16

Similarly, **cube numbers** form cubes. Don't believe me? Try it for yourself!

Exercise 6

1. List the first 10 **cube numbers**.

2. Express 160 as a product of its prime factors and write the answer in index form. (Note: index form and exponential form are the same.)

3. Calculate the value of:
   (a) 2^5 (b) 3^4 (c) 12^2

4. Fill in the missing number:
   (a) $5^{\square} \equiv 5 \times 5 \times 5 \times 5 \times 5 \times 5$

   (b) $1000 \equiv 10^{\square}$

5. Determine $2^5 \div 2^3$. Write your answer in index form.

Laws of Exponents

There are rules that govern how we multiply and divide exponents, which we shall discuss in this section.

Traditionally, if you were asked to evaluate $2^3 \times 2^5$, you may do so as follows:
$(2 \times 2 \times 2) \times (2 \times 2 \times 2 \times 2 \times 2) = 2 \times 2 \times 2 \times 2 \times 2 \times 2 \times 2 \times 2 = 2^8$

This answer is correct. But what if you were given $12^7 \times 12^{10}$? It would be a lot less fun to write all of these numbers out. Instead, mathematicians have developed rules which govern exponents and which make our calculations much easier and prettier to look at. (Yes, math can be pretty and many find it beautiful.)

1. **Product Rule:** $x^a \times x^b = x^{a+b}$
Example: $2^3 \times 2^5 = 2^{(3+5)} = 2^8$

2. **Quotient Rule:** $x^a \div x^b = \frac{x^a}{x^b} = x^{a-b}$
Example: $3^7 \div 3^5 = 3^{(7-5)} = 3^2$

3. **Zero Exponent Rule:** $x^0 = 1$
Example: $1,2345^0 = 1$

> There are a few other rules, but they are beyond the scope of this course. If you are interested in learning them, please see the Resources section of this guide.

Exercise 7

1. Verify the Product Rule and the Quotient Rule by working out the examples provided. That is, work them out "the long way".

2. Write as a power of 9:
(a) $9^7 \div 9^3$

(b) $\frac{9^2 \times 9^3}{9}$

Square Roots

Square roots are essentially the opposite of square numbers.

For example, the square of 4 (or 4²) is 16 so the square root of 16 is 4.

We normally write this as $\sqrt{16} = 4$, which reads "the square root of 16 equals 4".

Additional Examples
The square root of 25 is 5 (because 5² = 5 x 5 = 25)
The square root of 81 is 9 (because 9² = 9 x 9 = 81)

Remember, square roots are the opposite of square numbers. To truly be confident in finding a square root without a calculator, you must first be confident with square roots. I therefore encourage you to memorize the first 20 square roots.

An expectation of the syllabus is that you know square roots up to 400 without a calculator. Review the table below. You will be asked to recreate the table in Exercise 8.

Value	Square Root	Square Number
1	$\sqrt{1} = 1$	1²
4	$\sqrt{4} = 2$	2²
9	$\sqrt{9} = 3$	3²
16	$\sqrt{16} = 4$	4²
25	$\sqrt{25} = 5$	5²
36	$\sqrt{36} = 6$	6²
49	$\sqrt{49} = 7$	7²
64	$\sqrt{64} = 8$	8²
81	$\sqrt{81} = 9$	9²
100	$\sqrt{100} = 10$	10²
121	$\sqrt{121} = 11$	11²
144	$\sqrt{144} = 12$	12²
169	$\sqrt{169} = 13$	13²
196	$\sqrt{196} = 14$	14²
225	$\sqrt{225} = 15$	15²
256	$\sqrt{256} = 16$	16²
289	$\sqrt{289} = 17$	17²
324	$\sqrt{324} = 18$	18²
361	$\sqrt{361} = 19$	19²
400	$\sqrt{400} = 20$	20²

Surds are a special type of square roots. These are numbers which are left in square root form, like $\sqrt{3}$ or $\sqrt{2}$, because they cannot be simplified further. These numbers are **irrational**. If we tried to write them as decimals the digits would go on forever! In my experience, surds are not part of the BJC examination, but you may encounter them in later studies.

Exercise 8

1. Complete the table below:

Number	Value	Square Root
1^2		
2^2		
3^2		
4^2		
5^2		
6^2		
7^2		
8^2		
9^2		
10^2		
11^2		
12^2		
13^2		
14^2		
15^2		
16^2		
17^2		
18^2		
19^2		
20^2		

2. Play the square root game below.
https://www.mathgames.com/skill/8.6-square-roots-of-perfect-squares

3. Take the quiz below to see if you can recall the first 20 square roots in under 1 minute: https://www.sporcle.com/games/shs_mathteacher/square-roots

4. Determine if the below statement is true or false:

(a) $\sqrt{64} \geq 9$ 　　　　　　　　　(b) $\sqrt{9} \equiv 3$

Cube Roots

Cube roots are the opposite of cube numbers.

For example, the cube of 2 (or 2^3) is 8 so the cube root of 8 is 2.

We normally write this as $\sqrt[3]{8} = 2$, which reads "the cube root of 8 equals 2".

An expectation of the syllabus is that you know cube roots up to 1,000. Review the following table. You will be asked to recreate it in Exercise 9.

Value	Cube Root	Cube Number
1	$\sqrt[3]{1} = 1$	1^3
8	$\sqrt[3]{8} = 2$	2^3
27	$\sqrt[3]{27} = 3$	3^3
64	$\sqrt[3]{64} = 4$	4^3
125	$\sqrt[3]{125} = 5$	5^3
216	$\sqrt[3]{216} = 6$	6^3
343	$\sqrt[3]{343} = 7$	7^3
512	$\sqrt[3]{512} = 8$	8^3
729	$\sqrt[3]{729} = 9$	9^3
1000	$\sqrt[3]{1000} = 10$	10^3

Exercise 9

Complete the table below:

Cube Number	Value	Cube Root
1^3		
2^3		
3^3		
4^3		
5^3		
6^3		
7^3		
8^3		
9^3		
10^3		

Standard Index Form

Standard index form, also known as "scientific notation", is very useful for writing very large or very small positive numbers.

For example, 35,000,000,000,000 or 0.000000000000000015

We express numbers in standard form by multiplying by powers of 10.

For example, what is 3×10^5?
$3 \times 10^5 = 3 \times 10 \times 10 \times 10 \times 10 \times 10 = 300,000$

Q. What is $3 \div 10^2$?
$3 \div 10^2 = \dfrac{3}{10 \times 10} = \dfrac{3}{100} = 0.03$

Generally speaking, in **standard index form** numbers are written as follows:

$$a \times 10^n,$$

where a is a number between 0 and 10 (that is, 1, 2, 3, 4, 5, 6, 7, 8, 9) and n is an integer.

Because this guide concentrates on whole numbers, we shall focus writing only very large numbers in standard form.

Examples
Q. Write 7,000,000 in standard index form?
7,000,000 = **7 x 10⁶**
          6

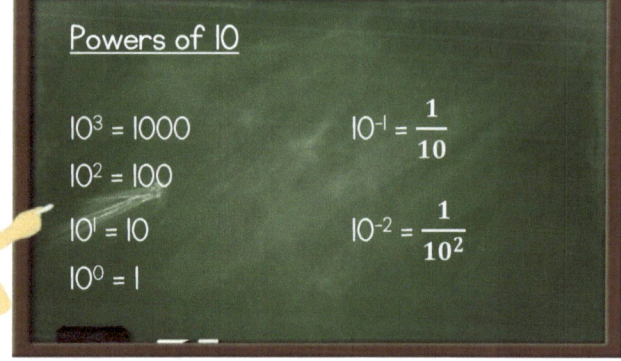

Q. Write 346,000 in scientific notation?
346,000 = **3.46 x 10⁵**
         5

We will explore standard index form further at the BGCSE level. Remember that standard index form and scientific notation are the same.

The Four Rules

This section will provide a revision of four basic number operations (+, −, ×, ÷) as applied to whole numbers. If any of the exercises are difficult for you, please review the materials listed in the Resources section of this guide.

Addition & Subtraction

Review the examples below then answer the questions that follow in Exercise 10.

Examples
Q. Calculate 3,263 + 2,103 + 104.

```
  1
  3623
  2103
+  104
  5830
```

Addition Notes:
1. Remember to always align the digits based on their place value: units on top of units, tens on top of tens, etc.
2. If the digits in a column add up to more than nine, the tens part is added to the next column.

Q. Calculate 9,034 - 4,105.

```
-4105
 4929
```

Subtraction Notes:
1. Remember to always align the digits based on their place value: units on top of units, tens on top of tens, etc.
2. If the number being subtracted is larger than the number it is being subtracted from, borrow from the next column and reduce the next column by 1.

Exercise 10

1. What is 325 + 357?

2. What is 9784 − 6785?

3. Add 16,662 and 14,397.

4. Karla has 50 guineps, 3 mangoes, 120 sea grapes and 4 guavas. How many fruits does she have in total?

Multiplication & Division

Review the examples below then answer the questions that follow in Exercise 11:

Examples

Q: What is 3,126 multiplied by 43?

```
      2
     11
   3126
 ×   43
 ─────
   9378
+125040
 ─────
 134418
```

Multiplication Notes:
1. Remember to always align the digits based on their place value: units on top of units, tens on top of tens, etc.
2. Add the result of the multiplication in order to get the final result.

If the exercises are hard for you, watch the videos below on YouTube:
- "Math Antics – Multi-Digit Multiplication Pt I"; and
- "Math Antics – Multi-Digit Multiplication Pt II".

Q: Solve 435 ÷ 5.

```
      87
   ┌─────
 5 │ 435
   - 40
   ─────
     035
    - 35
   ─────
      00
```

Division Notes:
Each digit below the line is divided by the number on the left.

If the exercises are hard for you, watch the videos below on YouTube:
- "Math Antics - Long Division with 2-Digit Divisors"; and
- "Math Antics - Long Division".

Exercise 11

1: What is 247 × 6?

2: What is 9784 ÷ 2?

3: Mr. Jones has a box of chips that he wants to share with his students. There are 49 bags of chips in the box and 15 students in his class. How many bags of chips does each student get? How many bags are remaining?

Order of Operations

In Mathematics, order of operations governs which operations (addition, subtraction, multiplication, division, exponents and grouping) take precedence over other operations.

There are many ways to remember the order of operations. A particularly common acronym is **PEDMAS** or "Please Excuse My Dear Aunt Sally":

P parenthesis
E exponents
D division
M multiplication
A addition
S subtraction

When applying the Order of Operations, we work from left to right.

Another useful acronym—though essentially the same thing—is **BODMAS**.

B brackets
O order (i.e. exponents)
D division
M multiplication
A addition
S subtraction

The acronyms simply state that brackets take precedence over all other operations, followed by powers or exponents and then division, multiplication, addition and subtraction, working from left to right. Let's apply the rules using examples.

Examples

1. Simplify $20 - (2 \times 6) + 3^3$

According to the rules, **brackets** or **parenthesis** are first.

$20 - (2 \times 6) + 3^3$
$= 20 - 12 + 3^3$ (Next, we evaluate the exponents.)
$= 20 - 12 + 27$ (Next, add or subtract, working from left to right.)
$= 8 + 27$
$= 35$

2. Simplify $4 + (2 \times 4^2 - 2) \div 5$

Step 1: Evaluate the exponents within the bracket → $4 + (2 \times 16 - 2) \div 5$
Step 2: Perform the multiplication within the bracket → $4 + (32 - 2) \div 5$
Step 3: Perform the subtraction within the bracket → $4 + 30 \div 5$
Step 4: Divide → $4 + 6$
Step 5: Add for the final answer → 10

3. $\dfrac{12 + 2 \times 4}{6 + 2^2}$

Step 1: Think of the problem as $(12 + 2 \times 4) \div (6 + 2^2)$
Step 2: We can work on both brackets simultaneously (at the same time) or one at a time → $(12 + 8) \div (6 + 4)$
Step 3: Continue to simplify the expressions in the brackets → $20 \div 10$
Step 4: Simplify for the final answer → 2

Note:
There are multiple ways to solve an order of operations problem. You may have decided to focus only on one set of brackets fully before working out the brackets I the denominator. Either way is fine as long as the rules are followed. That is, once you follow the order of operations, you cannot go wrong.

Exercise 12

1. $14 + 18 \div 2 \times 18 - 7$

2. $10 \div 5 + 10 - 9 \times 11$

3. $9 + 15 \div 5 \times 13$

4. $10 \times 12 - 14 \div 2 + 15$

Number Patterns & Sequences

A **sequence** is a set of numbers that are related in some way. The numbers are called the terms of the sequence.

Recognizing patterns and sequences is an important skill to master. The true aim is to train your eye to identify patterns and thus to improve your attention to detail. In the job market detail-oriented people and individuals who display attention to detail are highly sought after.

Here are a few sequences that are easy to recognize:

- **Even Numbers:** 2, 4, 6, 8, 10, 12...
- **Odd Numbers:** 1, 3, 5, 7, 9, 11...
- **Square Numbers:** 1, 4, 9, 16, 25, 36, 49, 64...
- **Cube Numbers:** 1, 8, 27, 64, 125...
- **Powers of 2:** 2, 4, 8, 16, 32, 64...
- **Powers of 10:** 10, 100, 1,000, 10,000, 100,000,...
- **Triangle Numbers:** 1, 3, 6, 10, 15, 21, 28,...

So far in this guide we have not discussed triangle numbers, so let us take a look at them below.

A triangle number sequence is a sequence generated from a pattern of dots which form a triangle. We learned that square numbers form a pattern of squares. Similar logic applies to **triangle numbers**.

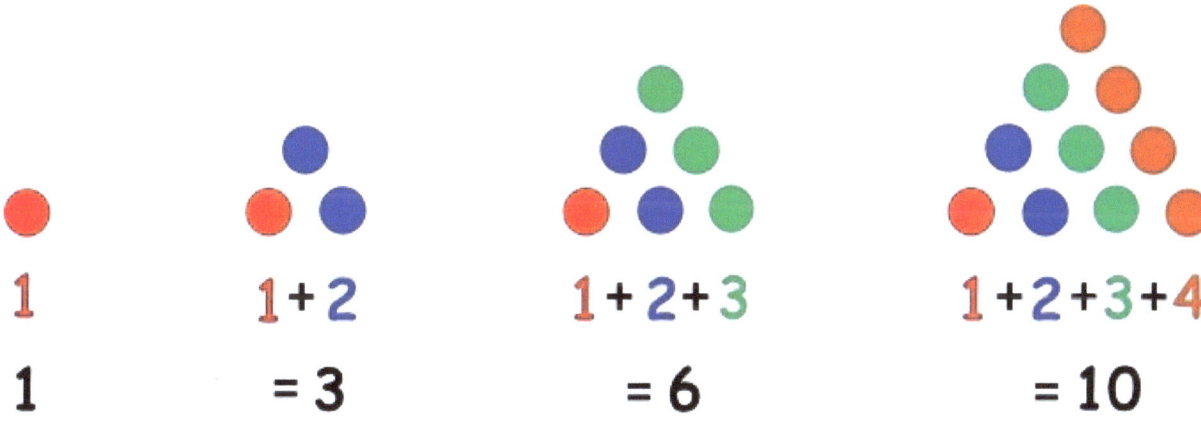

The rule applied to the triangular number sequence is:

$$\frac{n(n+1)}{2}$$

For example, to find the 50th term in a sequence of triangle numbers,

$$50^{th} \text{ term} = \frac{50(50+1)}{2} = 2550/2 = 1,275$$

Most sequences have rules such as "add 3" or "multiple by 2":

"add 3" sequence: 2, 5, 8, 11,....
"multiply by 2" sequence: 3, 6, 12, 24,....

It is up to you to identify the rule and to determine the additional or missing terms in a particular sequence or pattern.

A very famous sequence you may encounter on the exam is "**The Fibonacci Sequence**" which was created by Italian mathematician Leonardo Fibonacci (though some believe the sequence was developed earlier in India). In this sequence the rule is "add the two previous numbers together to get the next number":

Fibonacci Sequence: 1, 1, 2, 3, 5, 8, 13,....

The sequences you will encounter on the BJC exam are simple, linear sequences. The patterns should be easy to spot using trial and error and your knowledge of common sequences (squares, cubes, etc.). The questions that follow will help train your eyes to recognize patterns and sequences. Good luck!

Exercise 13

1: Based on the patterns below, draw pattern 4:

Pattern 1   Pattern 2   Pattern 3

2: Write down the next three numbers in each sequence. What is the rule for each sequence?

(a) 2, 4, 6, 8, 10; ____, ____, ____

(b) 3, 8, 13, 18, 23; ____, ____, ____

Exam Style Questions

1. Answer all questions in the spaces provided. Show all working.

(a)   5127
   + 3189
   ――――   [1]

(b)   6107
   − 809
   ――――   [1]

(c)   415
   ×   7
   ――――   [1]

(d) $3 \overline{) 3159}$ [1]

2. Write down the next two terms in the following sequence:

54; 53; 50; 45; 38; 29; _____; _____

3. Based on the hand of cards below, answer the following questions:

(a) Identify the square numbers, if any.   [2]

(b) Identify the prime numbers, if any.

4. (a) Express 196 as a product of its prime factors. [2]

(b) Write your answer in index form. [1]

(c) Calculate $\sqrt{196}$ [1]

5. Find the next two numbers in the sequences below.
(a) 7, 10, 13, 16, 19, _____, _____ [2]

(b) 200, 190, 181, 173, 166, _____, _____ [2]

6. The number of pencils in a pack is 12. James buys 3 packs of pencils. How many pencils does he have in total? [2]

7. A movie theater has a maximum capacity of 420. If there are 40 rows of seats in the theater, how many seats are in each row? [2]

8. Simplify $(1 + 7) - (2 \times 3) + (8 \div 4)$ [2]

9. Find the difference between terms for the sequence below and write down the next two terms in the sequence. [2]

6, 17, 28, 39, 50, _____, _____

10. List all prime numbers less than 30. [3]

11. Find the Highest Common Factor and Lowest Common Multiple of:
(a) 16 and 30 [2]

(b) 42 and 63 [2]

12. (a) 248 x 137 [1]

(b) 1356 – 349 [1]

13. The prime factors of 24, 30 and 45 are given below. [2]

20: 2 x 2 x 5
30: 2 x 3 x 5
150: 2 x 3 x 5 x 5

Use the factors to write down the HCF (Highest Common Factor) of the numbers.

14: Write down the value of a:

$$\sqrt{a} = 6$$

15: A mail man has to deliver 5 boxes. Each box weighs 2 pounds. If one of the boxes is taken away, calculate the weight of the remaining 4 boxes. [2]

16: Fill in the numbers in the square so that the totals are the same in all directions. [4]

15		16
	12	11
8		

17: (a) Evaluate $2^3 + 7^2$ [3]

(b) $(6,124 - 14) \div 5$ [2]

18: Here is a list of 8 numbers: 3, 5, 6, 11, 26, 7, 2, 8

From this list, write down:
  (a) 2 odd numbers [2]

  (b) 2 prime numbers [2]

  (c) a multiple of 13 [1]

  (d) a factor of 64 [1]

19. Write 856 000 000 in standard index form. [1]

20. 931 ÷ 7 [1]

21. Circle the number that is **NOT** a square number: [1]

   36   9   121   144   196   4

22. Circle a number that is **NOT** a prime number: [1]

   51   3   27   2   11

23. Find the value of $2^2 \times 3^3$. [3]

24. Simplify $2^5 \times 2^7$. Write the answer in exponent form. [3]

25.

64	2	18
7	12	11
8	144	25

From the numbers above, write down:

(a) a multiple of 4 [1]

(b) a number that is both square and cube [1]

(c) a multiple of 12 [1]

26. 8165 + 347 + 12 [1]

27. Write 4000 + 200 + 30 as an ordinary number. [1]

28. Use >, <, or = to make the statement true.

(a) 3^0 _____ 2 [1]

(b) 2 _____ $\sqrt[3]{8}$ [1]

(c) 4^2 _____ 4 + 4 [1]

29. Write the Highest Common Factor of 12 and 20. [3]

30. Evaluate $\sqrt{64} + 4^3$ [3]

31. Calculate the value of $2^3 \times 7$ [2]

32. Write down the numbers between 19 and 30 which have two factors only. [2]

33. Write down the next two terms in the sequence. [2]

$$1, 5, 9, 13, \underline{}, \underline{}$$

34. Calculate the value of $30 \div 5 + (2 \times 3) + 2^3$ [3]

35. In prime factor form: [2]

   $8 =$ $2 \times 2 \times 2$
   $28 =$ $2 \times 2 \times 7$

   What is the Lowest Common Multiple (LCM) of 8 and 28?

36. List all of the **natural numbers** that are greater than 3 but less than 10. [2]

37. Evaluate $6^2 + 4(3 - 1)$ [3]

38. Write as an ordinary number $5 \times 10^1 + 3 \times 10^2$ [3]

39. Determine the Highest Common Factor of the following numbers: [2]

   $48 =$ $2 \times 2 \times 2 \times 2 \times 3$
   $56 =$ $2 \times 2 \times 2 \times 7$

40. Write $600 + 30 + 2$ in standard form. [1]

41. List **ALL** one digit composite numbers. [2]

42. List
(a) a multiple of 3 between 31 and 37 [1]

(b) a prime number between 14 and 20 [1]

43. Simplify $72 \div 9 \times 3 \div 2$ [3]

44. Which is bigger **9065** or **9165**? [2]

45. Kevin says that 64 is a square number, not a cube number. Is he correct? Explain why or why not in full sentences. [2]

46. D'shanti made 372 cupcakes. 9 cupcakes fill a box.

(a) How many boxes will D'shanti fill with the cupcakes she baked? [2]

(b) How many cupcakes will be left over? [1]

47. Find the value of $6^3 - \sqrt{100}$ [3]

Standard Index Form

Standard index form, also known as "scientific notation", is very useful for writing very large or very small positive numbers.

For example, 35,000,000,000,000 or 0.000000000000000015

We express numbers in standard form by multiplying by powers of 10.

For example, what is 3×10^5?
$3 \times 10^5 = 3 \times 10 \times 10 \times 10 \times 10 \times 10 = 300,000$

Q. What is $3 \div 10^2$?
$3 \div 10^2 = \dfrac{3}{10 \times 10} = \dfrac{3}{100} = 0.03$

Generally speaking, in **standard index form** numbers are written as follows:

$$a \times 10^n,$$

where a is a number between 0 and 10 (that is, 1, 2, 3, 4, 5, 6, 7, 8, 9) and n is an integer.

Because this guide concentrates on whole numbers, we shall focus writing only very large numbers in standard form.

Examples
Q. Write 7,000,000 in standard index form?
7,000,000 = **7 x 10^6**
6

Q. Write 346,000 in scientific notation?
346,000 = **3.46 x 10^5**
5

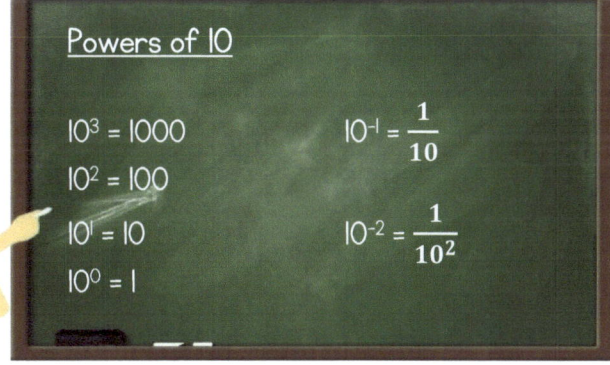

We will explore standard index form further at the BGCSE level. Remember that standard index form and scientific notation are the same.

Cube Roots

Cube roots are the opposite of cube numbers.

For example, the cube of 2 (or 2^3) is 8 so the cube root of 8 is 2.

We normally write this as $\sqrt[3]{8} = 2$, which reads "the cube root of 8 equals 2".

An expectation of the syllabus is that you know cube roots up to 1,000. Review the following table. You will be asked to recreate it in Exercise 9.

Value	Cube Root	Cube Number
1	$\sqrt[3]{1} = 1$	1^3
8	$\sqrt[3]{8} = 2$	2^3
27	$\sqrt[3]{27} = 3$	3^3
64	$\sqrt[3]{64} = 4$	4^3
125	$\sqrt[3]{125} = 5$	5^3
216	$\sqrt[3]{216} = 6$	6^3
343	$\sqrt[3]{343} = 7$	7^3
512	$\sqrt[3]{512} = 8$	8^3
729	$\sqrt[3]{729} = 9$	9^3
1000	$\sqrt[3]{1000} = 10$	10^3

Exercise 9

Complete the table below:

Cube Number	Value	Cube Root
1^3		
2^3		
3^3		
4^3		
5^3		
6^3		
7^3		
8^3		
9^3		
10^3		

Resources

Whole Numbers
http://www.bbc.co.uk/bitesize/ks3/maths/number/whole_numbers/revision/1/

Addition & Subtraction
http://www.bbc.co.uk/bitesize/ks2/maths/number/addition_subtraction/read/1/

Exponents and Roots
https://www.bbc.com/bitesize/guides/z66p34j/revision/1
https://www.lamission.edu/learningcenter/docs/mathlab/algebra/exponents_indices_powers_orders.pdf
http://www.bbc.co.uk/bitesize/ks3/maths/number/powers_roots/revision/2/

Square & Cube Numbers
https://www.bbc.com/bitesize/articles/z2ndsrd

Laws of Exponents
https://www.mcckc.edu/tutoring/docs/br/math/expon_logar/Exponent_Rules_Practice.pdf

Standard Form
https://www.teachitmaths.co.uk/attachments/27916/standard-form-poster.pdf
https://www.bbc.com/bitesize/guides/ztg987h/revision/1

Number Patterns & Sequences
http://www.bbc.co.uk/bitesize/ks3/maths/algebra/number_patterns/revision/1/
https://www.mathsisfun.com/algebra/triangular-numbers.html
http://www.cimt.org.uk/projects/mepres/allgcse/bkb12.pdf

www.ingramcontent.com/pod-product-compliance
Lightning Source LLC
Chambersburg PA
CBHW051935210526
45473CB00006B/2252